THE ARTISTIC SCROLLSAW

WONDERS OF NATURE

Terry Foltz

Schiffer Publishing Ltd®

4880 Lower Valley Road Atglen, Pennsylvania 19310

DEDICATION

To my uncle, Warren "Duke" Sawyer for getting me interested and started in scroll sawing and letting me use one of his saws. He is a wonderful craftsman and mentor.

Other Schiffer Books on Related Subjects

The Scroll Saw Book, Frank Pozsgai, $12.95
Scroll Saw Pictures, Frank Pozsgai, $12.95
79 Wildlife Patterns for the Scroll Saw, $14.95
72 Crafty Scrollsaw Patterns, Barbara G. McKee, $14.95

Designed by Mark David Bowyer
Type set in Bremen Bd BT / Zurich BT

ISBN: 978-0-7643-2676-9
Printed in China

Published by Schiffer Publishing Ltd.
4880 Lower Valley Road
Atglen, PA 19310
Phone: (610) 593-1777; Fax: (610) 593-2002
E-mail: Info@schifferbooks.com

For the largest selection of fine reference books on this and related subjects, please visit our web site at
www.schifferbooks.com
We are always looking for people to write books on new and related subjects. If you have an idea for a book please contact us at the above address.

This book may be purchased from the publisher.
Include $3.95 for shipping.
Please try your bookstore first.
You may write for a free catalog.

In Europe, Schiffer books are distributed by
Bushwood Books
6 Marksbury Ave.
Kew Gardens
Surrey TW9 4JF England
Phone: 44 (0) 20 8392-8585; Fax: 44 (0) 20 8392-9876
E-mail: info@bushwoodbooks.co.uk
Website: www.bushwoodbooks.co.uk
Free postage in the U.K., Europe; air mail at cost.

INTRODUCTION

This book has been written to walk you through the process of making a unique piece of artwork and to give you the tools to prepare and design your own patterns. Entitled "Fish On," the project is a fly fishing scene with Mount Saint Helens in the background. It is a local mountain and everyone that lives in Washington State is familiar with it and knows what it looks like. What makes the scene exceptional is the number of trees and the oversized trout coming out of the water.

I strive to put as much detail in my work as possible. This is one thing that sets my work apart from other artists or craftspeople. This will usually get people's attention when they view my work at art shows. You will see this when you look at the patterns and photos of my artwork.

You have probably heard the old saying "It's easier said than done," and it's true. Once you start adding the extra detail it will be more difficult and take longer to saw, but you will really like the results. When you begin the piece it will seem as if it will take forever to complete it. After you have done it for a while, it will get easier and will not take as long. Hopefully you will tough it out and not short cut your piece by eliminating the extra detail.

The pieces I make range from twenty-four inches to fifty inches in length and twelve inches high. My favorite wood is big leaf spalded maple. I utilize the wood grain and coloration to give my art work a three-dimensional look. You will see this in some of the photos in the gallery at the back of the book.

Generally my wood thickness will be .375 inches, however, it will vary according to the type of wood and/or grain. Unlike most people I use a chain saw to remove any straight lines in the wood. It gives my art work shape and a rustic look. Most people will never notice the chain saw marks unless they are shown to them.

This book should give you a better understanding of how to design and make your own patterns. Making your own patterns or designs is fairly easy and is very rewarding. The simplest way to do this is to take parts of patterns, copy them and cut them out. Once you have the extra pieces of different patterns, you can modify them and make them your own patterns. A simple example would be to remove the mountain in my pattern and replace it with another mountain that may be located in your area or one you have seen or photographed on a trip. If you do this, try to use distinctive mountains that people will recognize. It will really enhance you work. If I find a mountain that I like in a magazine or a photo, I will trace it and put in what I call cut lines to give the mountain its contour lines. Once this is done you can copy, enlarge, or reduce it to fit your pattern. The same can be done with trees or any other object you may want to use. It is amazing what you can create. When you finish designing your pattern, make a copy of it and save it in a file. If the pattern turns out to be something you like, you may want to use it again. Having to design the same pattern the second time is rather frustrating...I know this from past experience. I have a filing system set up for all of my finished patterns. I also have a filing system for trees, mountains, sunsets, barns, and all the animals I use in my artwork.

Good luck on making this piece and I hope you will benefit from this book.

INSTRUCTION AND TIPS

Choosing the right blade for your work is fairly easy, especially if you have been sawing for a while. If you are like me, you will probably prefer sawing one type of pattern and wood thickness. If this is the case, you will probably end up with one or two blades that you will use 95% of the time. The blade guide that is shown will give you the information you need to make an educated decision as to the saw blade that is best for you.

There is no exact science to blade selection. The type of wood, wood thickness, saw speed, complexity of design, and feed rate or pressure on the blade will determine the type of blade that is best for you. The end result will be what feels good to you or what someone else has recommended.

That is what I will do as we start cutting out the pattern. My advice is to look at the saw blade chart and make your decision based on the information in it and the recommendation of someone else that you know.

Hawk Woodworking describes the six types of blades in their brochure from which the following was extracted and adapted with permission:

1. Regular Skip Tooth Blades:
Skip tooth blades are exactly that. Some teeth are skipped on the blade. This allows for clearing more material from the kerf while cutting. These are great blades to use if you are a beginner to scroll sawing.

2. Skip Tooth Reverse Blades:
Skip Tooth Reverse Blades are the next step up from regular Skip Tooth Blades. They have several teeth at the bottom of the blade that face in the opposite direction to the main down-cutting teeth. This will eliminate splinters on the bottom of the workpiece that you are cutting. If eliminating as much sanding as possible is your goal, this is the type of blade you should be using.

3. Double Skip Tooth Reverse Blades
Double Skip Tooth Reverse Blades have two teeth together followed by a flat space for efficient chip removal. There are also reverse teeth to help eliminate splinters on the bottom of your workpiece. These blades cut faster than the previous two and leave clean edges in wood and plastic.

4. Premium Speed Ground Tooth Reverse Blades:
"Ground Tooth" are the keys word with this style of blade. These blades are pretty much the same as Skip Tooth Reverse Blades, but with small teeth that have been ground to shape rather than simply filed. These blades are much sharper, cut in an extremely straight line, and leave a very smooth surface. These are great blades that cut in an aggressive manner, but not recommended for the beginner.

5. Bi-Directional Cut Blades:
Bi-Directional Cut Blades have a unique tooth design that cuts on both the up and down strokes.

The Hawk Premium Scroll Saw Blades

SIZE	ITEM #	MAXIMUM THICKNESS	TYPE OF PATTERN	TEETH/ INCH	WIDTH	BLADE THICKNESS
Skip Tooth Blades Top-grade high-speed fretsaw with widely-spread teeth, hardened, 5 inch long, teeth set.						
2/0	866-7000	1/4″ Thick	Intricate	27	.022″	.010″
2	866-7002	3/8″ Thick	Intricate	22.5	.030″	.013″
3	866-7003	1/2″ Thick	Detailed	20	.031″	.013″
5	866-7005	1/2″ Thick	Average	17	.037″	.015″
7	866-7007	1″ Thick	Average	15	.043″	.017″
9	866-7009	1 3/4″ Thick	Simple	13	.053″	.018″
12	866-7012	2 1/2″ Thick	Simple	9.5	.067″	.020″
Skip Tooth Reverse Blades Top-grade high-speed fretsaw, hardened, with reverse teeth, 5 inch long. Splinter free finish on top and bottom.						
2/0	866-7100	1/4″ Thick	Intricate	15	.027″	.008″
1	866-7102	3/8″ Thick	Intricate	15	.030″	.011″
3	866-7103	1/2″ Thick	Detailed	12	.035″	.013″
5	866-7105	1/2″ Thick	Average	12	.037″	.015″
7	866-7107	1″ Thick	Average	12	.043″	.017″
9	866-7109	1 3/4″ Thick	Simple	12	.050″	.018″
12	866-7112	2 1/2″ Thick	Simple	8	.067″	.020″
Double Skip Tooth Reverse Blades Top-grade high-speed fretsaw. Double teeth, hardened, with reverse teeth, 5 inch long. Splinter free finish on top and bottom.						
2/0	866-7200	1/4″ Thick	Intricate	23	.027″	.008″
1	866-7202	3/8″ Thick	Intricate	20	.030″	.011″
3	866-7203	1/2″ Thick	Detailed	20	.035″	.013″
5	866-7205	1/2″ Thick	Average	17	.037″	.015″
7	866-7207	1″ Thick	Average	15	.043″	.017″
9	866-7209	1 3/4″ Thick	Simple	11.5	.050″	.018″
12	866-7212	2 1/2″ Thick	Simple	8	.067″	.020″

SIZE	ITEM #	MAXIMUM THICKNESS	TYPE OF PATTERN	TEETH/ INCH	WIDTH	BLADE THICKNESS
Premium Speed Cutter/Ground Tooth Reverse Blades Top Grade high-speed fretsaw, hardened, with reverse teeth. 5″ long. Splinter free finish on top and bottom.						
2/0	866-7300	1/4″ Thick	Intricate	15	.027″	.008″
1	866-7302	3/8″ Thick	Intricate	13	.030″	.011″
3	866-7303	1/2″ Thick	Detailed	13	.035″	.013″
5	866-7305	1/2″ Thick	Average	13	.037″	.015″
7	866-7307	1″ Thick	Average	10	.043″	.017″
9	866-7309	1 3/4″ Thick	Simple	8	.050″	.018″
12	866-7312	2 1/2″ Thick	Simple	7	.067″	.020″
Bi Directional Cut Blades Top-grade high-speed fretsaw, hardened, each 3rd tooth reversed. 5″ lg. Splinter free finish on top and bottom.						
2/0	866-7500	1/4″ Thick	Intricate	15	.027″	.008″
1	866-7502	3/8″ Thick	Intricate	15	.030″	.011″
3	866-7503	1/2″ Thick	Detailed	13	.035″	.013″
5	866-7505	1/2″ Thick	Average	13	.037″	.015″
7	866-7507	1″ Thick	Average	10	.043″	.017″
9	866-7509	1 3/4″ Thick	Simple	8	.050″	.018″
12	866-7512	2 1/2″ Thick	Simple	8	.067″	.020″
Spiral Blade Reverse Top-grade fretsaw for wood. With reverse teeth and splinter free finish on top and bottom surface.						
2/0	866-7400	1/4″ Thick	Intricate	10	.027″	.008″
1	866-7402	3/8″ Thick	Detailed	7	.030″	.011″
5	866-7405	1″ Thick	Average	7	.037″	.015″

***Please note:** There is no mix and match with these new blades to make up your gross amount.

Because of this two-way cutting action, this type of blade allows for cutting with more control, while leaving a smooth, splinter-free finish. The blade can be put in either way, so there is no upside-down with these blades. They will cut a little slower than other blades, but are good for plastic and Plexiglas®. May cause burn on harder, thicker stock.

6. Spiral Blades:
Spiral Blades are twisted in a spiral fashion. This allows the teeth to cut on all sides of the blade. It is not necessary to turn your workpiece when using a spiral tooth blade as all sides of the blade will cut. This can be an advantage when cutting large workpieces, where it is impossible to turn the workpiece around without hitting the support arm of the saw. Spiral blades tend to leave a rough, wide surface and cannot make a tight or sharp corner. It's also very difficult to follow a straight line.

The life of the blade will depend on you. Correct tension on the blade will increase its life. My advice would be to turn your saw down to its slowest speed and saw a few pieces at that speed. Your blade will last longer, primarily due to the slower speed. This is because the slower speed generates less heat and less heat leads to longer blade life.

But the biggest, most important advantage of slower speeds is more accurate sawing. It will make a craftsman out of an average scroll sawyer. For some people this will be difficult; however, if you commit to this technique, you will make fewer mistakes and your work will be more accurate. If it takes you a third longer to cut out a piece but you only have to do it once, it will be much more rewarding and you will generate pieces that look better.

One additional advantage is that it will train you to have patience. This is a very desirable trait when you are doing a large complex piece. When you try this, the slowest speed of your saw may, in fact, be too slow. If so, turn up the speed to the slowest speed you can be comfortable with.

Whichever blade you believe will work best for you is the one you should choose. Different people will use different blades when cutting the same piece. My favorite blade is the Hawk skip tooth 2/0 blade. I like it because of the high number of teeth per inch and the thickness of it. The Hawk blade seems to be stronger and more consistent in tooth pattern and it also seems to last longer. The only draw back is the price; they are approximately thirty percent higher than other blades. Olson, Great American, and Pegas are also very good blades; I have used them all. There are other blades on the market that I am sure are just as good, but I have not tried any of them.

GENERAL INFORMATION

Most of you already have a magnification light. If not, you will need to purchase one before you begin sawing this pattern. With the detail in this pattern, it is necessary. If you are sawing slowly and are using a magnification light, you will be able to correct a mistake before it happens. What I am saying is that you will be able to see what you are doing and have the time to fix things before they go wrong.

Sawing always generates a fair amount of fine dust. If you saw any for length of time, you will need to use a filter of some kind. You can use a mask, a vacuum system, or a filter system that you can purchase. Another option is simply to make a filter system using a box fan and a furnace filter. It is a simple solution that I read about in the *Scroll Saw Woodworking and Crafts* magazine some time ago. If you do not already have one, you can purchase a box fan for less than twenty dollars and a good, high quality filter for less than ten dollars. The filter should be the same size as the box fan. Tape or tie the filter to the suction side of the fan and locate it close to your work area. Turn the fan on while you are working. It will suck the dust particles into the filter and they will collect on it. It is not fancy, but it works. When the filter looks like it is full, use a vacuum cleaner to remove the dust and use it again.

Scroll Saw Woodworking and Crafts is a wonderful source for information and new ideas. Below is an chart of Possible Reactions to Woods which they published as part of an article on "Wood Toxicity." It has very important information to help you make educated decisions on breathing protection, and is used with permission.

WOOD TOXICITY

Courtesy of *Scroll Saw Woodworking and Crafts* magazine, and published in their Summer 2005 edtion.

Wood	Class (irritant or sensitizer)	Reaction Type	Potency	Source	Incidence
Alder	Irritant*	Respiratory, eye & skin	No info***	Dust	No info
Ash	Irritant	Respiratory	No info	Dust	No info
Avodire	Irritant*	Respiratory, eye & skin	No info	Dust	No info
Baldcypress	Sensitizer**	Respiratory	Small	Dust	Rare
Beech	Sensitizer	Respiratory	Great	Dust	Rare
Birch	Sensitizer	Respiratory, Nausea	Great	Dust	Rare
Black locust	Irritant	Nausea	Great	Dust	Rare
Bubinga	Irritant	Eye & skin	No info	Dust	No info
Red cedar, Eastern	Irritant	Respiratory, eye & skin	No info	Dust	Common
Red cedar, Western	Sensitizer	Respiratory	Great	Dust, leaves & bark	Common
Cocobolo	Irritant	Respiratory, eye & skin	Great	Dust & wood	Common
Ebony	Irritant & Sensitizer	Respiratory, eye & skin	Great	Dust & wood	Common
Elm	Irritant	Eye & skin	Small	Dust	Rare
Goncalo alves	Sensitizer	Eye & skin	Small	Dust & wood	Rare
Greenheart	Sensitizer	Respiratory, eye & skin	Extreme	Dust & wood	Common
Ipe	Irritant	Respiratory, eye & skin	No info	No info	No info
Mahogany	Irritant	Respiratory, eye & skin	Small	Dust	Rare
Maple (usually only spalted)	Irritant	Respiratory, eye & skin	Great	Dust	Rare
Oak, red	Irritant	Nasal	Great	Dust	Rare
Padauk	Irritant	Respiratory, eye, skin, & nausea	Extreme	Dust & wood	Common
Purpleheart	Sensitizer	Eye, skin, & nausea	Small	Dust & wood	Rare
Rosewood	Irritant & Sensitizer	Respiratory, eye & skin	Extreme	Dust & wood	Common
Sassafras	Sensitizer	Respiratory, nausea, & nasal cancer	Small	Dust & wood	Rare
Teak	Sensitizer	Eye & skin	Extreme	Dust	Common
Walnut, black	Sensitizer	Eye & skin	Great	Leaves & bark	Common
Willow	Sensitizer	Nasal cancer	Great	Dust	Common

***An irritant** causes and immediate reaction each time the wood is used.
****A sensitizer** does not necessarily irritate, but it makes a person more likely to be strongly affecte by a wood classed as an irritant. If you are exposed to an irritant after being exposed to a sensitizer, you are more likely to have a more serious reaction to the irritant.
*****No info** indicates that the information for this wood is still being developed.

SOURCING YOUR MATERIALS

Wood can be found at any lumberyard or hardware store; therefore, wood in general is very easy to find. On the other hand, unique spalded or burl wood is not that easy to find. Those of us who live in the Northwest are lucky and have more access to this type of wood. The key is knowing where and how to look for it. Magazines such as Scroll Saw Woodworking and Crafts have advertisements that will show various vendors of wood. The internet is also an excellent source. If you live in a rural area that has a small saw mill, it may be another good source for wood. Most of the wood discarded at a mill is wood that will work great for the artwork I do. I have a couple of small sawmills that save wood for me. They are glad to get rid of it and I occasionally will make a piece for them; generally that is all the pay they will want. Listed below are names of suppliers. Some I have used and some I have not. Most will ship to you and some will have the capability to e-mail you photos of the wood and some have web sites you can access.

Another source of wood is to join a wood working club or a wood carvers club. If they have shows in your local area you can visit them and probably find out where they get their wood. If you do not see any clubs listed in the phone book, you might go to a craft show. They usually have people who are there selling their work. This may be your best source. The last is to find a specialty woodworking store. If you are lucky and have a good store near you, the people working there will be a valuable resource. They may sell all types of exotic and unique wood or they should be able to locate the type of wood you are seeking.

Specialty Wood Sources

Woodcrafters
601 6th Street
Portland OR
800-777-0511

Randle Woods
P.O. Box 96
Randle, WA 98377
360-497-2071
e-mail: woodshop@randlewoods.com

Got Wood
www.thehandsomewoodman.com
e-mail: jackjudy@a-z.net
360-260-6908

Hardwood Components
e-mail: hardwood@earthlink.net
800-859-2144

JX3 Hardwoods, LLC
e-mail: jx3hardwoods.earthlink.net

Woodshed Gifts and Clocks
1729 Shelby Street
Indianapolis, IN 46203
800-388-6322
Buckeye wood slabs

Catalogs

Listed below are the names, phone number and websites of a number of catalogs that sell all types of supplies for scroll saws, including patterns, books, saw blades, and other accessories for the avid scroller. The most useful items are saw blades. They sell them by the gross at a big discount, versus buying them at your local hardware store in packages of twelve. Note: Once you order something from them you will get their catalog for life. If you like scroll sawing, I guess that isn't so bad.

Wildwood Designs
1-800-470-9090
www.wildwooddesigns.com

Cherry Tree
1-800-848-4363
www.cherrytreetoys.com

The Berry Basket
1-800-206-9009
www.berrybasket.com

The Wooden Teddy Bear
1-888-762-9149
www.woodenteddybearonline.com

The Winfield Collection
1-800-946-3435
www.winfieldcollection.com

Scroller LTD
1-800-486-6053
www.scrollerltd.com

I am sure there are other catalogs, but these are the only ones that I regularly receive.

THE PROJECT "FISH ON"

Choosing the right piece of wood for a project, I believe, is an art in itself, or at least that is what I like to think. When you have chosen the right piece, shape, and color, it will transform an average project into a stunning piece of artwork. People at art shows tell me all the time that the wood I use is beautiful. A gorgeous piece of wood is half the beauty of your artwork; therefore you should take some time to select a nice piece for yourself. If you have a good source for wood, pick out something you think will have some color or grain. This will influence the piece you are getting ready to cut. The size of the wood we will use in this piece is approximately twenty-four inches in length by twelve inches high.

The piece of wood that I have selected is western big leaf maple. It has a nice grain and will have a red tint to it when it is finished with linseed oil. As you can see, I have shaped the wood to give it a natural look and to allow my pattern to fit the wood.

I first clamp the wood to my planer to secure it. Next I take my chain saw and shape the wood. Another way to shape it is to use a band saw or your scroll saw. The key is to eliminate any straight lines. They seem to take away from the artistic beauty of the work.

Notice that I have used the chain saw again to rough up all the edges of the wood, giving it a rustic look. When I originally designed this pattern the wood I chose already had this shape. It just seems to fit this scene. It also gives the piece a unique look, so it does not look like I just bought it at the local lumberyard.

After I have finished with the chain saw, I plane the wood to the thickness I want. For this piece it is approximately .375 inches thick (3/8"). This is not an exact dimension; none of my work has the same thickness. When planning maple or other hardwoods, you will need to take small cuts out of the wood. If you are in a hurry and take too large a cut, you will probably start taking chips out of the wood. I prefer my pieces to be fairly thin. I believe it gives them more of a three dimensional look. If the wood is too thick it will look blocky and it is harder to cut with the scroll saw. It also will break more of your saw blades.

Next you will want to sand your wood to the finish you desire. For those of you without a planer, you should start your sanding at a lower grit. Unless you have rough-cut wood, you can start sanding with 100 or 150 grit paper. As you can see in the photo, I only use an orbital sander to finish my wood.

Next you will attach your pattern to the wood. Notice the pattern fits the contour of the wood. You usually design your layout to fit the wood, but since I have made this piece before, I have cut the wood to fit the pattern.

I start with 225 grit paper, go to 400 grit, and for the finishing touch use 600 grit. You can get by with 225 grit, but you will be happier if you use the 400 grit. That is really all you will need to get a very smooth finish. The 600 grit is used to take off some of the rough edges. It is like sharpening your knife on a stone. Once you get it just right, you rub it with a leather strap to set the finish. It also gives your wood more shine in the grain lines. When people touch and feel your piece, they will be impressed with the smoothness of the wood.

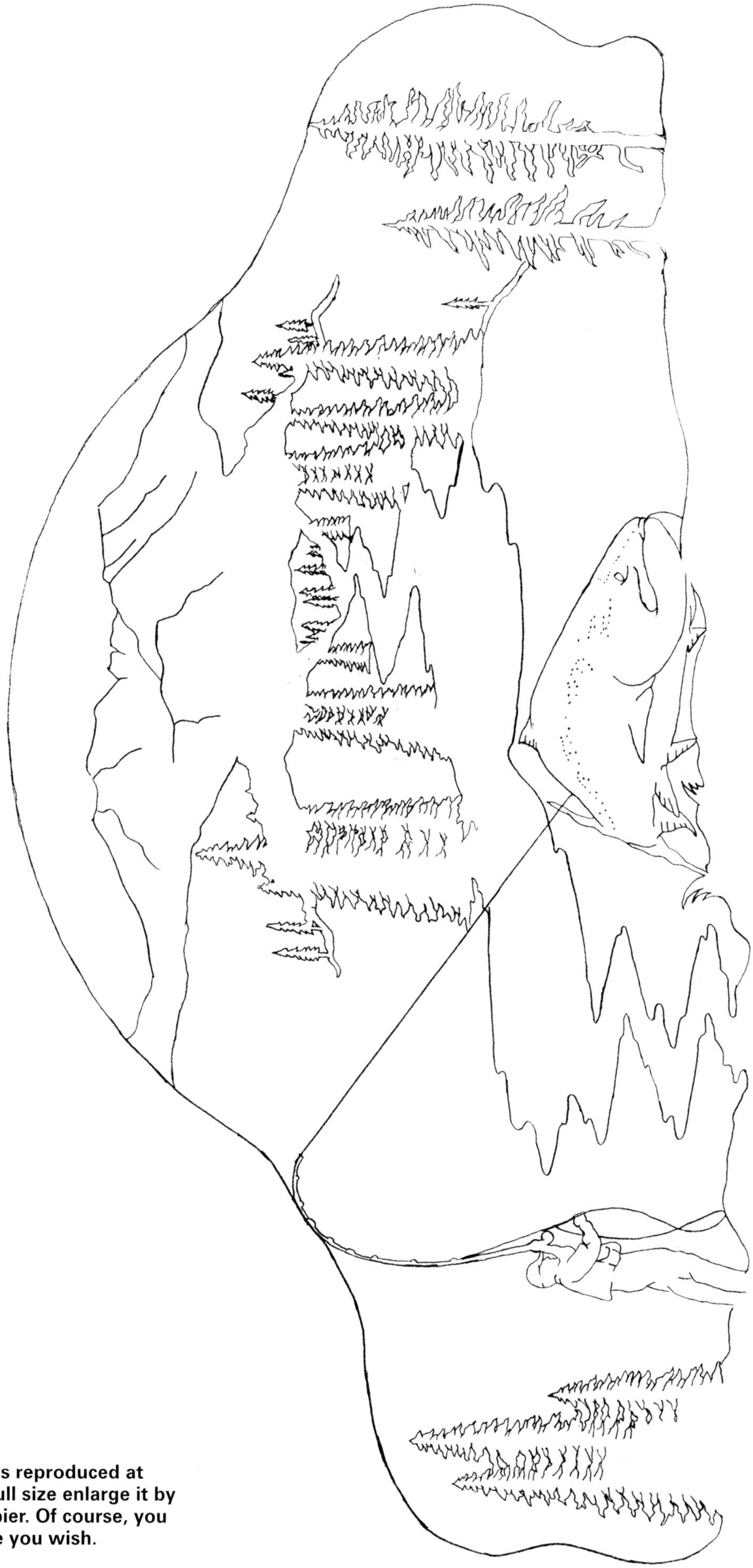

The project pattern is reproduced at 60%. To bring it to full size enlarge it by 167% on a photocopier. Of course, you can use it at any size you wish.

If you will look closely at the center of the pattern, you will see that I have used transparent tape on the top and bottom to hold the pattern in place. I center and level the pattern to the exact position I want it, then tape it in place. By doing this, the pattern doesn't move when gluing it to the wood.

I fold the pattern in half at the location of the tape.

Next, I place rubberized glue on the backside of the pattern, then secure to the wood. There are other ways to secure your pattern, but this works best for me. The main objective is to keep the pattern secured in position while gluing. This is very important. One of the worst things that can happen is to cut out a pattern and then realize it is not level on the wood. If it is not, start over again.

Once the pattern is secured to the wood and you have finished cutting, the paper remnants are fairly easily removed. Later we will discuss what to do if you have trouble removing the pattern.

Next I drill out the holes for the saw blade. I use a drill press to drill the holes. When you are drilling holes very close together, you will want your holes to be vertical. The only way to do this is with a drill press. Since I make numerous intricate cuts in my patterns, I use a 1/32-inch drill bit. You can order them from one of the catalogs listed earlier. The 1/32 drill bit is fairly fragile. If you aren't extremely careful with them, they will break. If you order them, be sure to order at least ten, as you will probably break a few during the learning curve!

If you look closely at the drill press, you will see a wire gauge drill chuck attached to the chuck. You will need this special attachment. The chuck on a regular drill press will not hold a 1/32 inch bit straight or very tight. The attachment only costs between six and ten dollars. But the bits and miniature chuck attachment at the same time.

We are now set to drill. I use the same bit for all holes. Generally about a third of the holes require this small size bit, but instead of changing back and forth, I use one bit for all the drilling. Drill all the interior areas of the pattern. There are approximately eighty drill holes in this pattern. Take your time. This can be tedious work and the bit is fragile.

PREPARING TO SAW THE PATTERN

First let's look at a couple of things that you will need to cut out this pattern. You may or may not have all of these, but we will go over all of them nonetheless.

In this photo there are two items you will need for this project. The first is a pair of reading glasses that have a two power magnification or larger. The next item is the magnification light. You will need this for extra lighting and clarity for sawing out the pattern. I cannot stress this enough. The larger you can make your pattern appear when sawing, the easier it will be to accurately saw the pattern. One mistake in the wrong place can be a disaster. The light and magnification will turn an average sawyer into a craftsman. You will be able to see yourself starting to make a mistake before it is too late.

Hopefully you can see the effect of good lighting in the photo above. When you begin to insert the saw blade into the drill hole, move the magnification light in front of your wood. The light will shine down on your saw platform and reflect up on the wood allowing you to see the drill hole much more easily, especially the 1/32-inch drill holes. Try it with and without the light and you will see the difference.

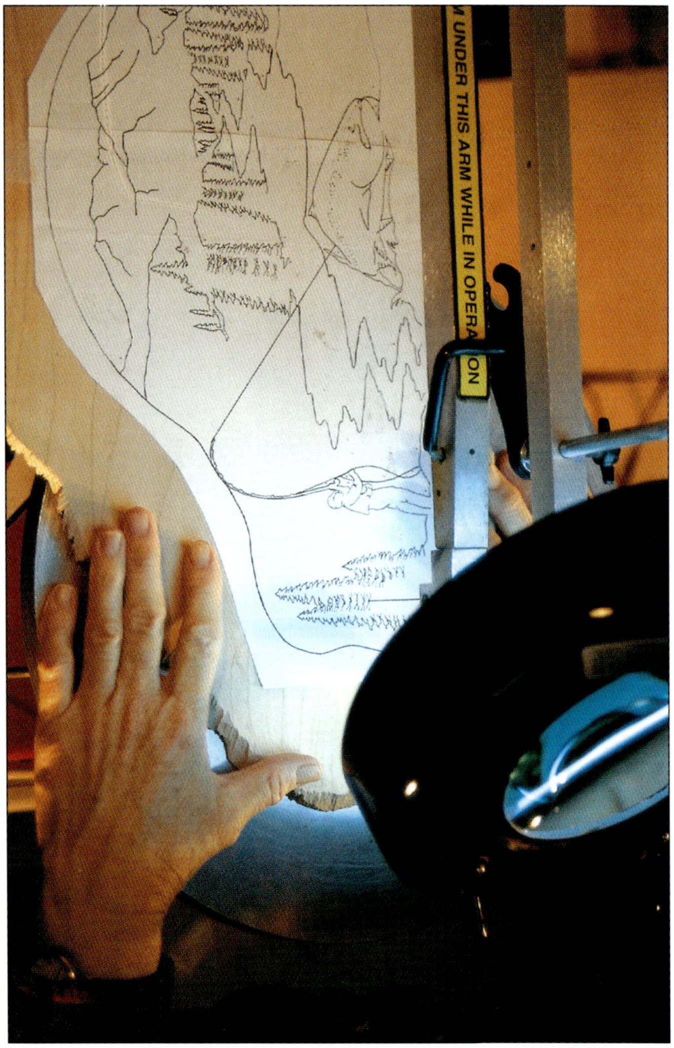

I generally start from one side or the other. This time I have started on the left side, cutting out the detail of the trees. There are sixteen drill holes in the three trees.

THE SAWING PROCESS

There is a preferred sequence to what you want to saw out first. I always start by sawing out the small, detailed portions of the pattern. In this piece I start with the inside of the trees.

As I mentioned earlier, I use a 2/0 skip tooth blade with twenty-seven teeth per inch. It gives me plenty of cutting power to make acute cuts in the wood. Be sure to set your saw speed at the lowest speed. This is critical in making small minute cuts. Let the saw do its work. Do not try to force the blade. All that will do is cause you to make a mistake and break more blades. One thing I do is to tighten the tension of my blade tighter than normal. It gives me the extra tolerance when cutting fine detailed work. This does reduce blade life, but it is worth it in the long run.

As you can see, I have continued to cut the minor detail out of the trees and have come back to the left side and stared cutting out larger portions. This is the way I usually cut out my work. I cut the smallest detail first, then move up in size.

I have also cut the outside tree detail of the three trees to the left of the fisherman. I have only cut the trees, but have left a solid base for cutting out the fly rod, etc. I will come back later and cut out the outside portion.

In this photo you can see I have cut the larger areas of the trees. I am continuing the detail portion of the pattern. The easy cuts will come later. I do all the hard work first, because there is less chance of breakage and it keeps me from rushing the process. I am removing the excess wood from the pattern with a round pick. It is small and strong enough to remove the cutout piece of wood.

Occasionally the sawed out portion of wood is a little hard to remove. It is usually a result of not sawing vertically due to the thin blade and/or too much pressure on the saw blade. This is another reason we do the inside cuts first. You have solid wood all around you so you can push harder on the wood to remove it.

Now I move to the other side of the pattern and cut out the tallest tree on the right side.

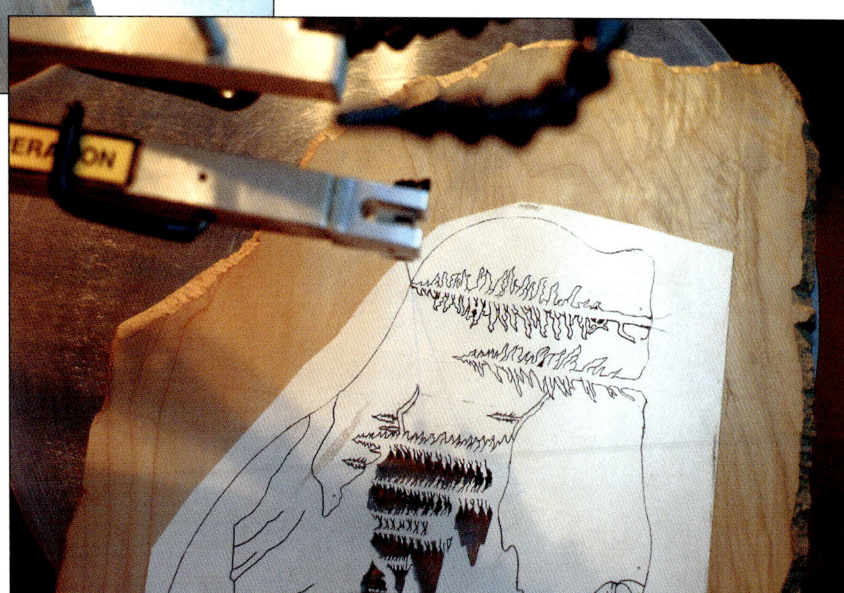

I started on the left side of the tree first, beginning at the bottom and stopping at the top of the tree.

I have the tree secured at the top for support.

I remove the saw blade and start cutting the other side of the tree.

Doing it this way gives support to the tree 100 percent of the time while cutting it out.

Next I cut out the trees to the left of the one I just finished.

I start at the bottom on the right side...

and continue to cut the outside portion of the trees until I get to the far right, at the base of the mountain as shown in the photo.

Progress.

With the trees completed, I cut out the bottom of the trout and its detail.

I cut out the fins, mouth detail, and the eye.

I have already cut out the tail of the fish and remaining piece of wood.

Progress.

Progress.

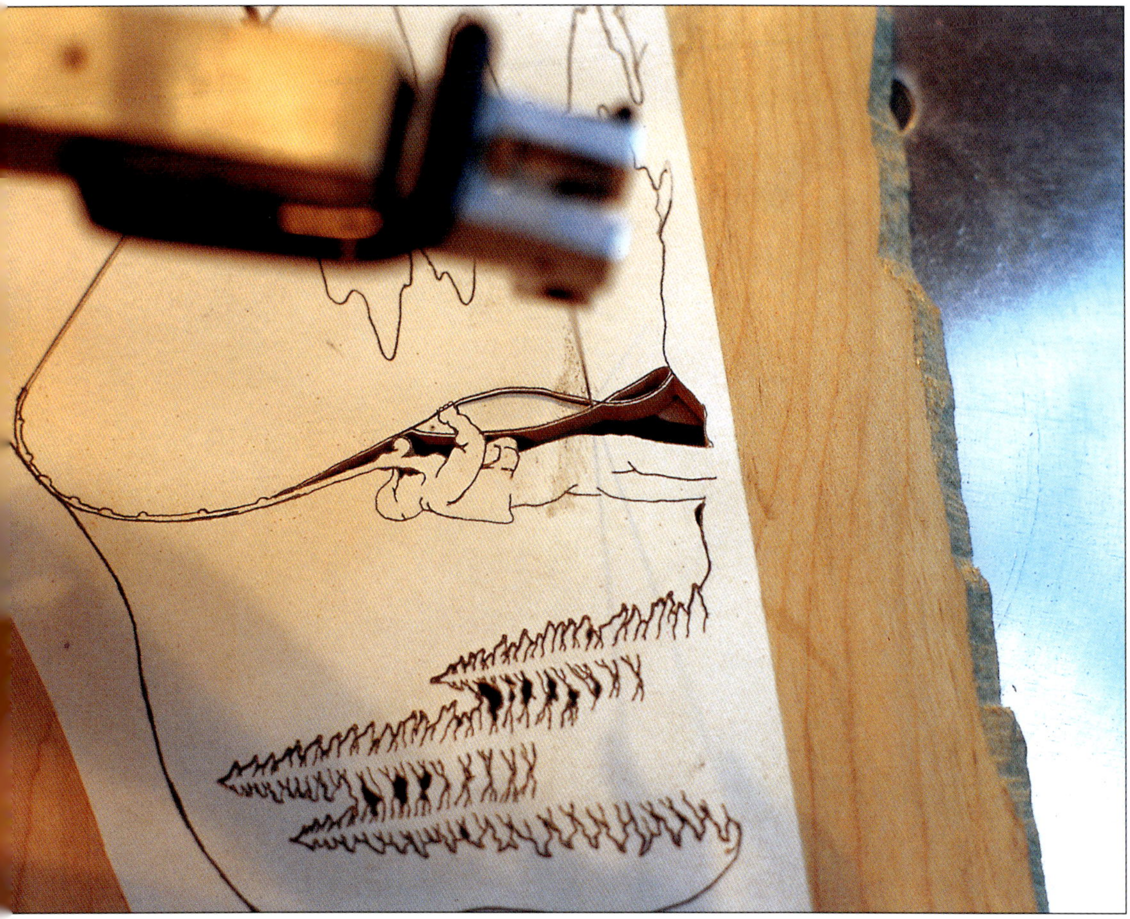

Now I cut out portions that require a steady hand. When you do this, especially the first time, take it very slow. The fishing line will be very fine. Here I cut just outside the pattern line. You want to have a little white showing that will give you approximately 1/16 inch for the line. If you do not feel comfortable doing this, take a straight edge and enlarge the fly line, then cut it the way you always do.

If you look closely you will see that I am cutting out the right side of the fly rod and the line going to the fish. I do this side first so I can cut out the eyelets on the fishing rod. When you cut out the eyelets, you do not want the wood to flex and possibly break. All the wood on the left side of the rod is stationary. This is the key to doing fine detailed work. When I cut the other side of the fly rod I will be cutting in a straight line, putting less pressure on the wood.

Next I will cut out the left side of the trees. I prefer to start from the bottom. I only cut up to the top of the trees.

I remove the saw blade and reinsert the blade in the hole just above the fly rod. What I am doing here is the same as I did before. I am cutting down the fly line having a solid base to cut from.

The next step is to cut out the outside portion of the trees and the bottom of Mt. St. Helens.

The reslt with the wood removed.

Now I cut out the area around the trees and the fly fisherman. I started at the base of the fisherman and will saw up to the top of the fly rod.

At that point I reverse direction and saw the outside area of the pattern down to the base of the trees, completing the cut.

Now I go back over and finish cutting out the left side of the trees on the right of the pattern.

I am almost finished. All that is left to do is Mount St. Helens. I will start on the left side cutting the mountain and contour lines, then finish cutting out the top. The cutting is now complete.

With all the excess wood removed, it is now time to remove the remaining portion of the pattern.

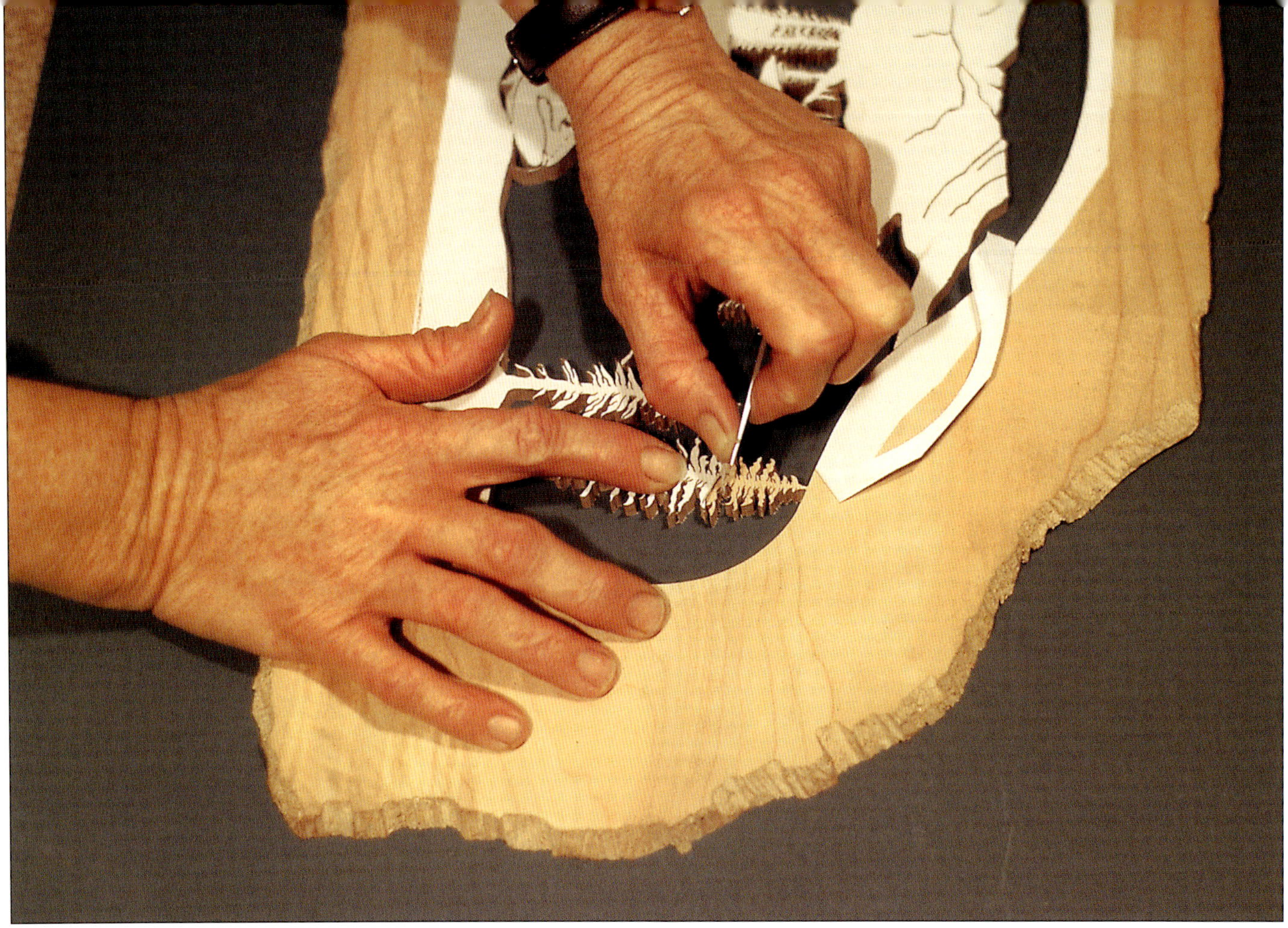

Removal of the pattern can be easy or, at times, be very slow and delicate. I use a small pocketknife to help in the removal of the paper. I start at the top of the tree on the right side.

With my hand I use the knife to lift up the paper and pull down while I use my other hand to hold the wood down against the table. This gives the tree all the support it needs while removing the pattern.

I will continue this process across the pattern until it is removed.

With the pattern removed, there is glue residue left on the wood. It has to be removed. Where there is excess glue, I rub it off with my finger.

Once this is done I sand the wood to move the remaining glue and to acquire the finish I want. I will start out again with 225 grit on the front and back of the piece.

Next I will use 400 grit...

and, finally, 600 grit paper on the front only.

Now I will remove all the frayed or rough edges on the backside of the wood with a small file. After this has been completed, all I have to do is blow and wipe off the excess dust.

This portion is optional. What I do here is put the spots on the trout with my woodburning tool. If you do not have this tool you can do this with a pencil, a Dremel or other rotary tool, the end of a knife, or anything with a sharp point. What I do is place the pattern back over the trout and burn in the spots through the paper. You can do the same thing with any of these tools.

Doing this adds more realism to the trout. I hope you will try it.

Finally it is time to finish the piece. I use Tried and True polymerized linseed oil to bring out the true colors of the wood and seal it.

As you can see, I use a medium artist's paintbrush to put on the oil. You will want to use a fairly small brush in order to go inside all the cutout areas and to put a complete coat of oil on the wood. I generally put on two coats of oil. Other finishes will also work. You may have your favorite. If not, try the linseed oil.

The Fish On scroll saw art piece is now complete.

THE GALLERY

The Linskof's

The Meginness's

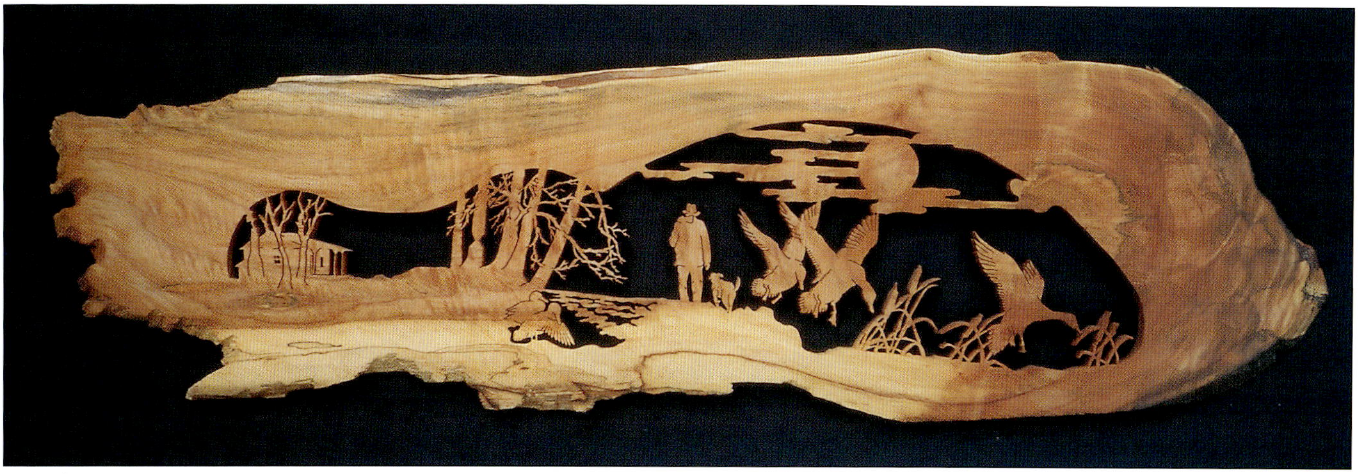

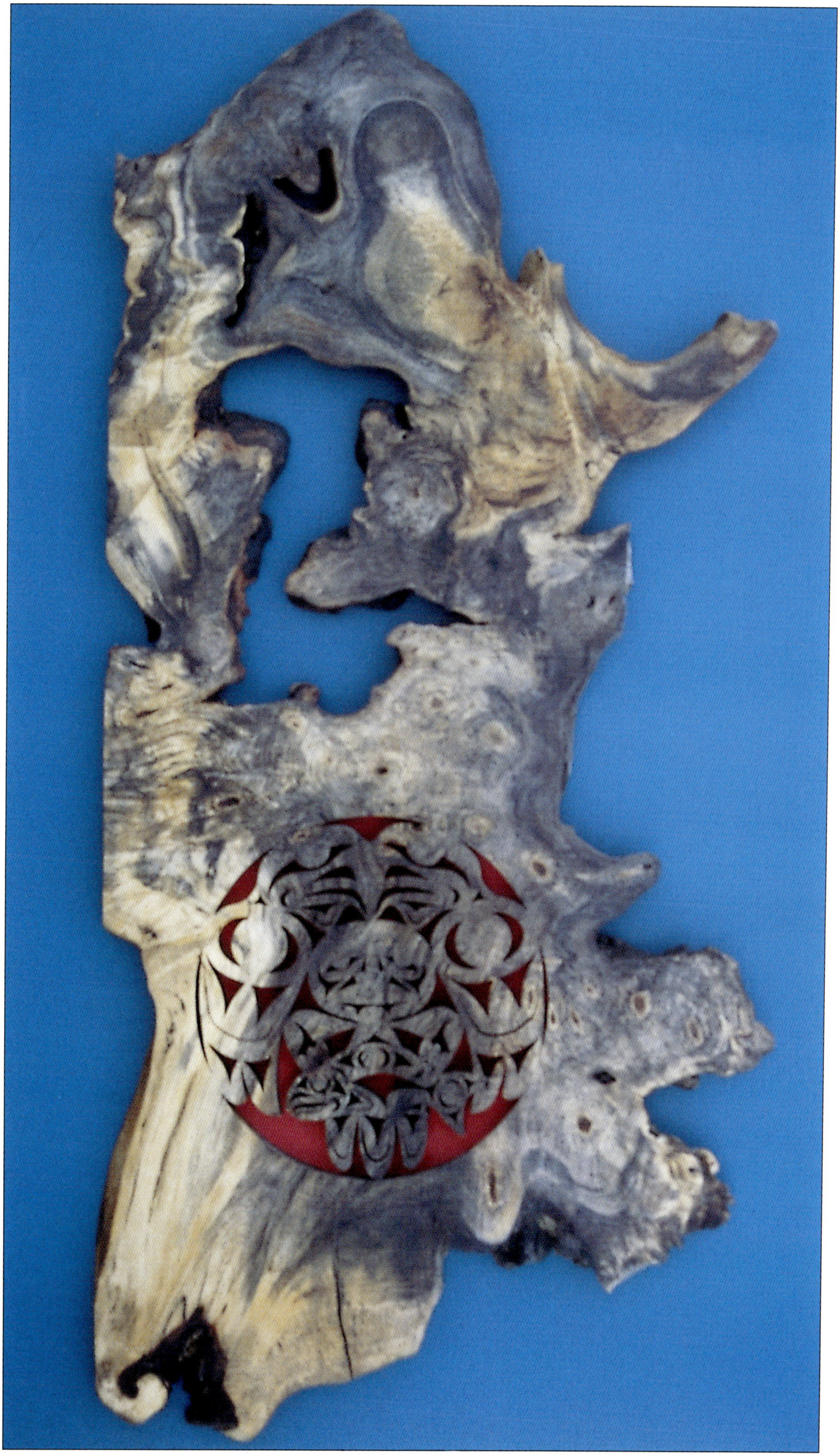

OTHER PATTERNS

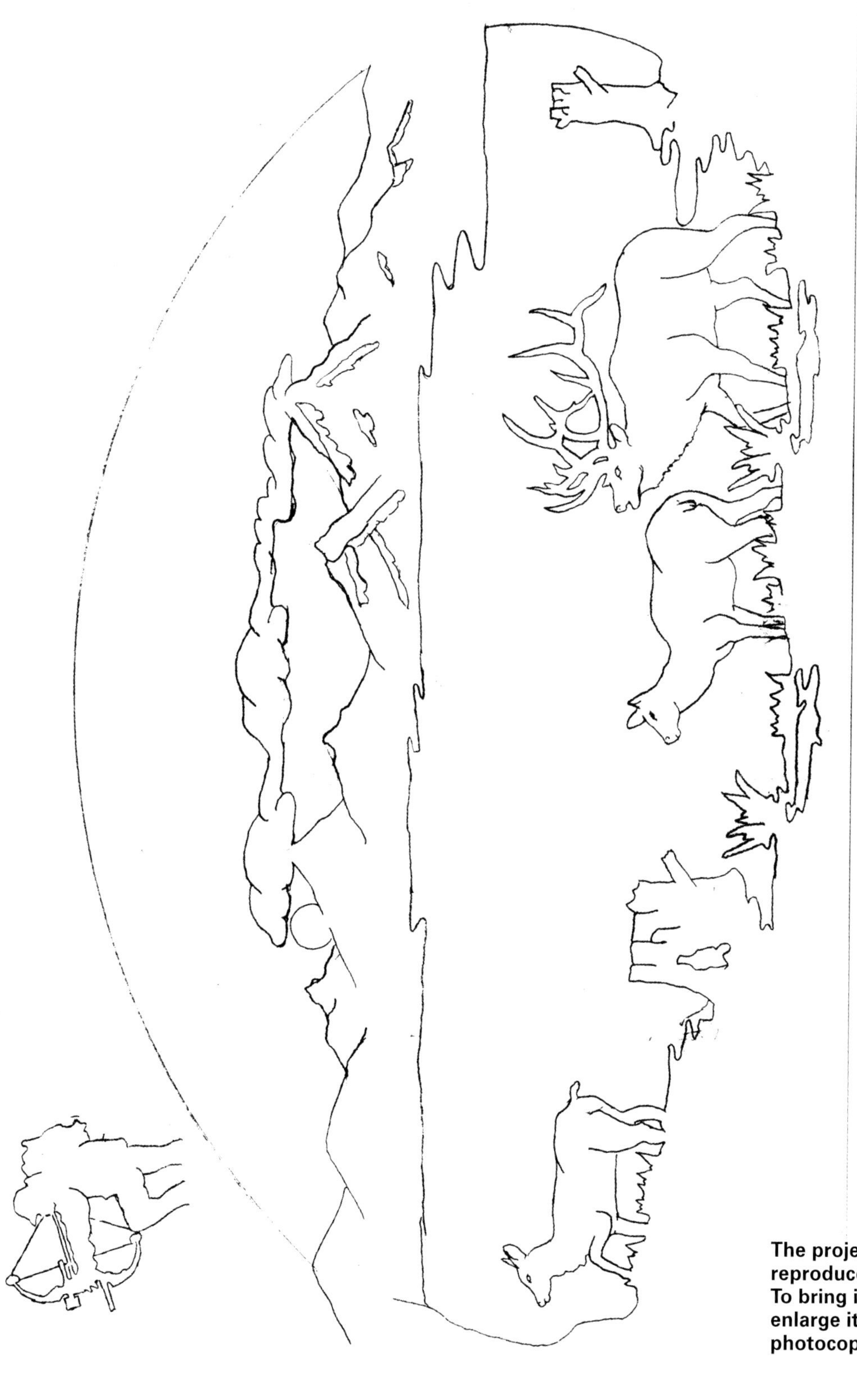

The project pattern is reproduced at 75%.
To bring it to full size enlarge it by 133% on a photocopier.

The project pattern is reproduced at 75%. To bring it to full size enlarge it by 133% on a photocopier.

The project pattern is reproduced at 100%.

The project pattern is reproduced at 90%. To bring it to full size enlarge it by 111% on a photocopier.

The project pattern is reproduced at 100%.

Both project patterns are reproduced at 75%. To bring it to full size enlarge it by 133% on a photocopier.